ANIMALS AT RISK

Tigers

by Rachel Grack

BLASTOFF! READERS 2

BELLWETHER MEDIA • MINNEAPOLIS, MN

Blastoff! Readers are carefully developed by literacy experts to build reading stamina and move students toward fluency by combining standards-based content with developmentally appropriate text.

Level 1 provides the most support through repetition of high-frequency words, light text, predictable sentence patterns, and strong visual support.

Level 2 offers early readers a bit more challenge through varied sentences, increased text load, and text-supportive special features.

Level 3 advances early-fluent readers toward fluency through increased text load, less reliance on photos, advancing concepts, longer sentences, and more complex special features.

★ **Blastoff! Universe**

This edition first published in 2022 by Bellwether Media, Inc.

Library of Congress Cataloging-in-Publication Data

Names: Koestler-Grack, Rachel A., 1973- author.
Title: Tigers / by Rachel Grack.
Description: Minneapolis, MN : Bellwether Media, 2022. | Series: Animals at risk | Includes bibliographical references and index. | Audience: Ages 5-8 | Audience: Grades 2-3 | Summary: "Simple text and full-color photography introduce beginning readers to the threats to and protections of tigers. Developed by literacy experts for students in kindergarten through third grade"-- Provided by publisher.
Identifiers: LCCN 2021046026 (print) | LCCN 2021046027 (ebook) | ISBN 9781644875926 (library binding) | ISBN 9781648346033 (ebook)
Subjects: LCSH: Tiger--Juvenile literature. | Tiger--Behavior--Juvenile literature. | Tiger--Conservation--Juvenile literature.
Classification: LCC QL737.C23 K645 2022 (print) | LCC QL737.C23 (ebook) | DDC 599.756--dc23
LC record available at https://lccn.loc.gov/2021046026
LC ebook record available at https://lccn.loc.gov/2021046027

Editor: Kieran Downs Designer: Brittany McIntosh

Printed in the United States of America, North Mankato, MN.

Table of Contents

Top Cats

Tigers are big cats with striped fur. They live in parts of southern and eastern Asia.

They are powerful hunters.

Sumatran tiger

There were once nine **subspecies** of tigers. Three types are **extinct**.

The other six are **endangered**. People have been their greatest enemy.

Tiger Range

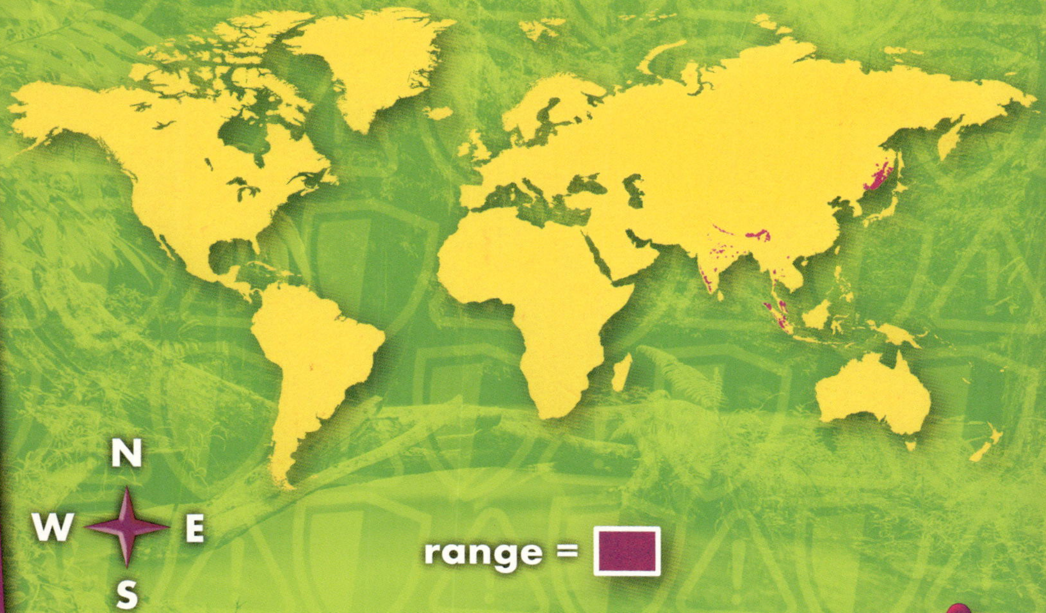

In Danger!

Poachers hunt tigers for their fur and body parts.

Tiger bones are sometimes used in **traditional** medicines. Some people even eat tiger meat.

tiger fur

Tiger Stats

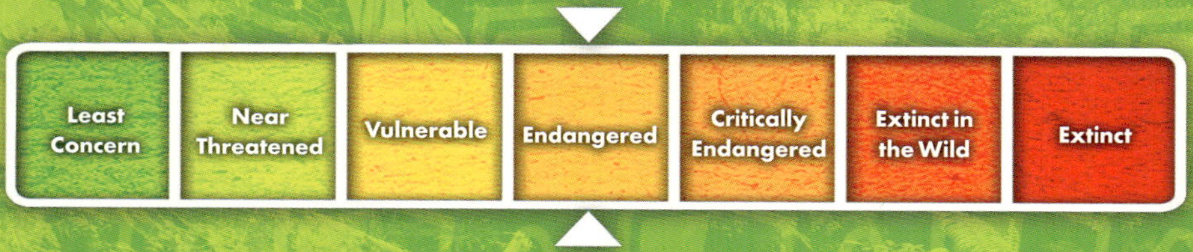

conservation status: endangered

life span: up to 20 years

Tigers have lost most of their **home ranges**. People cleared land for farms and roads.

Climate change also destroys tiger **habitats**.

Threats

1

people need farmland

2

people cut down trees

3

tigers have less food and fewer places to live

Save the Tigers!

Tigers keep their **ecosystems** healthy. The ecosystems give both people and animals fresh food and water.

Without tigers, people and animals would suffer.

The World with Tigers

1

more tigers

2

healthy forests

3

more food and fresh water for people and animals

Governments pass **laws** to help tigers. They ban medicines made from tiger parts.

Indochinese tiger

Rangers patrol tiger habitats and stop poachers.

Tigers need large home ranges. Some governments pay people to move away from tigers.

Farmers learn new ways to farm on less land.

Countries set aside land just for tigers. Pathways between **reserves** bring tigers together.

Bridges and underpasses let tigers travel safely around roads.

Slowing climate change helps tigers. Using less **energy** can help slow climate change.

Adopting a tiger is another way to save these big cats!

Glossary

adopting—taking over the care for someone or something; people who adopt wild animals give money for someone else to care for them.

climate change—a human-caused change in Earth's weather due to warming temperatures

ecosystems—communities of plants and animals living in certain places

endangered—in danger of dying out

energy—power to make things work

extinct—completely gone in the wild

habitats—the places where animals live

home ranges—lands on which animals or groups of animals live, hunt, and travel

laws—rules that must be followed

poachers—hunters who catch or harm animals illegally

rangers—people in charge of protecting an area of land

reserves—lands set aside for wildlife

subspecies—groups of one type of animal that live in different parts of the world

traditional—related to customs, ideas, or beliefs passed down from one generation to the next

To Learn More

AT THE LIBRARY

Carney, Elizabeth. *Wild Cats.* Washington, D.C.: National Geographic, 2017.

Hansen, Grace. *Siberian Tigers.* Minneapolis, Minn.: Abdo Kids Jumbo, 2019.

Murray, Julie. *Tigers.* Minneapolis, Minn.: ABDO Publishing, 2020.

ON THE WEB

FACTSURFER

Factsurfer.com gives you a safe, fun way to find more information.

1. Go to www.factsurfer.com.
2. Enter "tigers" into the search box and click .
3. Select your book cover to see a list of related content.

Index

The images in this book are reproduced through the courtesy of: Eric Isselee, front cover, p. 3; Matjoe, front cover (top back), pp. 3, 23; Elena Anisimova, front cover (bottom back); RAJU SONI, p. 4; slowmotiongli, pp. 5, 16; Benjamin B, p. 6; IgorCheri, p. 8; Ondrej Prosicky, pp. 8-9; Subi Sridharan, p. 10; Rich Carey, p. 11 (top left, top right); ATTOMY, p. 11 (bottom); enciktepstudio, p. 12; Archna Singh, p. 13 (top left); AJP, p. 13 (top right); Dmytro Gilitukha, p. 13 (bottom); TigerStocks, p. 14; Talukdar David, p. 15; Baby2M, p. 17; Sukpaiboonwat, p. 18; Sourabh Bharti, p. 19; Igor Anfinogentov, p. 20; Freder, p. 21; Stepan Kapl, p. 23.